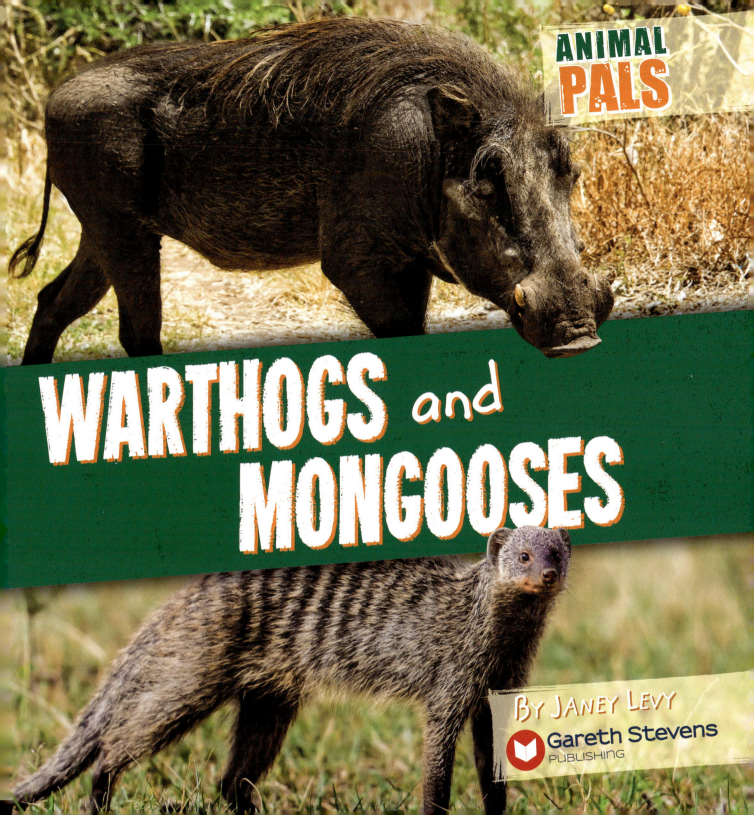

Please visit our website, www.garethstevens.com. For a free color catalog of all our high-quality books, call toll free 1-800-542-2595 or fax 1-877-542-2596.

Library of Congress Cataloging-in-Publication Data

Names: Levy, Janey, author.
Title: Warthogs and mongooses / Janey Levy.
Description: New York : Gareth Stevens Publishing, [2022] | Series: Animal pals | Includes index.
Identifiers: LCCN 2020038030 (print) | LCCN 2020038031 (ebook) | ISBN 9781538266977 (library binding) | ISBN 9781538266953 (paperback) | ISBN 9781538266960 (set) | ISBN 9781538266984 (ebook)
Subjects: LCSH: Warthog–Juvenile literature. | Mongooses–Juvenile literature. | Mutualism (Biology)–Juvenile literature.
Classification: LCC QL737.U58 L485 2022 (print) | LCC QL737.U58 (ebook) | DDC 599.63/3–dc23
LC record available at https://lccn.loc.gov/2020038030
LC ebook record available at https://lccn.loc.gov/2020038031

First Edition

Published in 2022 by
Gareth Stevens Publishing
29 E. 21st Street
New York, NY 10010

Copyright © 2022 Gareth Stevens Publishing

Designer: Andrea Davison-Bartolotta
Editor: Monika Davies

Photo credits: Cover (top), p. 1 (top) Mark Newman/The Image Bank/Getty Images; p. 5 (top) Ayzenstayn/Moment/Getty Images; p. 5 (bottom) Ryan Jack/500px/Getty Images; p. 5 (inset) Rainer Lesniewski/Shutterstock.com; pp. 7 (map), 9 (map) Peter Hermes Furian/Shutterstock.com; p. 7 (inset) Gallo Images-Anothy Bannister/DigitalVision/Getty Images; p. 9 (inset) Nimit Virdi/500px/Getty Images; p. 11 Ashour Rehana/Moment/Getty Images; p. 13 Rini Kools/Shutterstock.com; p. 15 SoopySue/E+/Getty Images; p. 17 Anup Shah/Corbis Documentary/Getty Images Plus/Getty Images; p. 18 Danielle Carstens/Shutterstock.com; p. 19 (top) Martin Harvey/Gallo Images/Getty Images Plus/Getty Images; p. 19 (bottom) Richard Packwood/Oxford Scientific/Getty Images Plus/Getty Images; p. 20 EcoPic/iStock/Getty Images Plus/Getty Images; p. 21 Chris Burt/Shutterstock.com.

All rights reserved. No part of this book may be reproduced in any form without permission in writing from the publisher, except by a reviewer.

Printed in the United States of America

CPSIA compliance information: Batch #CSGS22: For further information contact Gareth Stevens, New York, New York at 1-800-542-2595.

CONTENTS

Mammalian Mates 4
Wild Warthogs 6
More About Mongooses 8
Warthogs' Parasite Problem 10
Warthog Wallow 12
Warthogs and Mongooses 14
Why Warthogs and Mongooses Are Special 16
Warthogs and Birds 18
Warthogs and Turtles 20
Glossary 22
For More Information 23
Index 24

Words in the glossary appear in **bold** type the first time they are used in the text.

MAMMALIAN MATES

Warthogs and banded mongooses are both African **mammals**, but they are very different creatures. Warthogs are large, piglike mammals. They might look scary, but they are herbivores, or plant eaters. Banded mongooses are much smaller in comparison. They look somewhat like cats or weasels and are carnivores, or meat eaters.

So, what's the connection between these two animals? Surprisingly, they have a **relationship** that benefits them both! You'll discover more about these mammals and their relationship inside this book.

FACT FINDER!

A mutualistic relationship is a relationship between two different species, or kinds, of animals that benefits both of them.

Warthogs and banded mongooses both live south of the Sahara Desert in Africa.

WILD WARTHOGS

Warthogs belong to the same family as pigs. But a warthog would never be mistaken for a pig. Big, sharp **tusks** stick out of a warthog's mouth. Their huge head is covered with the "**warts**" that give them their name. Warthogs are mostly bald, but they have a mane running down the middle of their back.

Warthogs usually search for food during the day. At night, they sleep in warm **burrows**. The burrows give them a place to raise their young and hide from predators.

FACT FINDER

The "warts" on a warthog's face aren't true warts. They are actually thick pads of skin that **protect** warthogs when they fight.

This map shows the areas in Africa where warthogs live. Male warthogs are large animals that can weigh up to 330 pounds (150 kg)!

Africa

Where Warthogs Live

MORE ABOUT MONGOOSES

There are over 30 species of mongooses. Banded mongooses are a species that gets its name from the dark bands running up and down across their back. These bands set them apart from all other mongoose species.

Banded mongooses are social animals and like to live in groups. Their groups can have 10, 20, or even more members. They hunt for food during the day. Bugs are their favorite food, but they also eat eggs, fruit, birds, snakes—and even rats. Yum!

FACT FINDER!

Termites in Africa build huge mounds above their home. Banded mongooses often like to make their dens in old termite mounds.

This map shows where banded mongooses live in Africa. Banded mongooses are small animals that weigh around 3 to 5 pounds (1.4 to 2.3 kg).

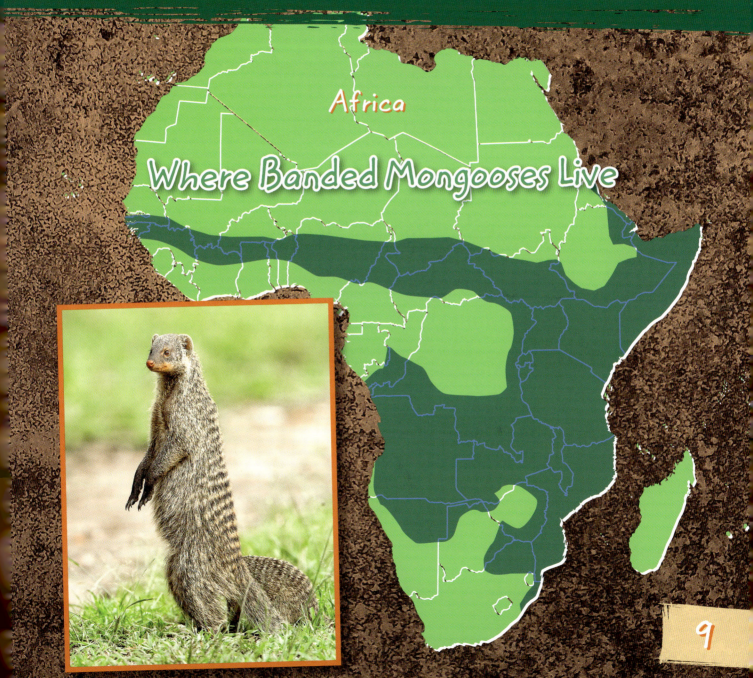

Africa

Where Banded Mongooses Live

WARTHOGS' PARASITE PROBLEM

Warthogs are smart animals. They usually like to look for food in the early morning or evening. But they'll change to being active at night if they live where humans hunt them.

But still, **parasites** give them problems. All sorts of parasites bother warthogs. If you have a dog or cat, you might be familiar with two of their parasites: ticks and **fleas**. Tsetse flies are also a problem for warthogs. It's hard for warthogs to deal with these parasites on their own.

FACT FINDER!

Ticks lie in wait in the burrows where warthogs go to find safety. That's pretty sneaky!

You might be surprised to learn these large animals can run up to 30 miles (48 km) per hour!

WARTHOG WALLOW

If you had ticks or fleas, you'd want to get rid of them, right? Warthogs do too. They have a trick for removing parasites. Warthogs like to wallow, or roll around, in mud. The mud dries, trapping the parasites. Warthogs can then rub the mud—and parasites—off their body.

But warthogs live in areas with a dry season. During these times, water is hard to find. Without water, there's no mud. Warthogs must have another way to get rid of parasites.

FACT FINDER

You need to drink water every day to stay healthy. But during the dry season, warthogs can go months without water!

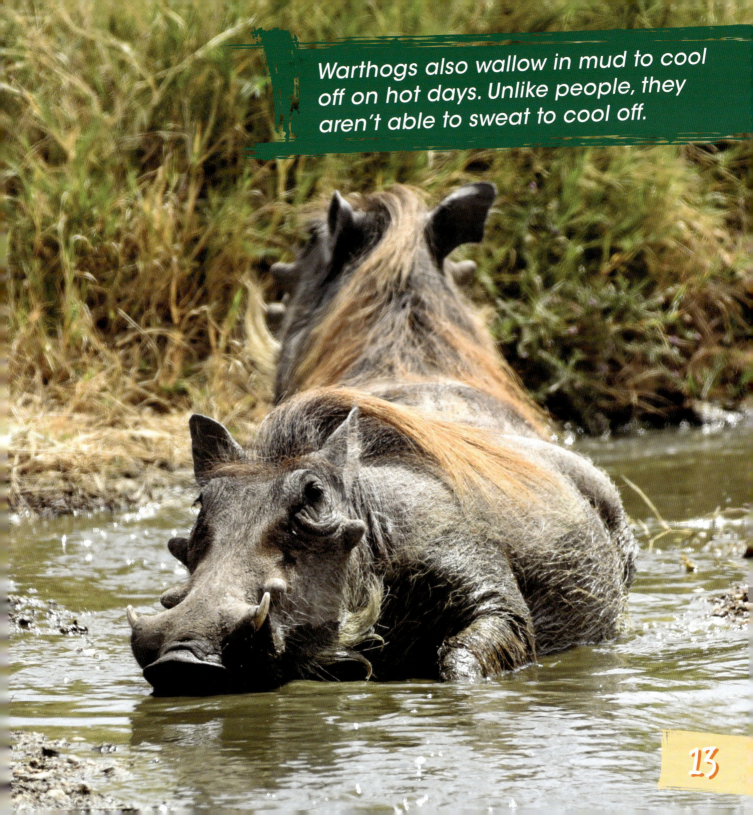

Warthogs also wallow in mud to cool off on hot days. Unlike people, they aren't able to sweat to cool off.

WARTHOGS AND MONGOOSES

In the African country of Uganda, warthogs have found a creature to help them with their parasite problem. When a warthog sees a group of banded mongooses, it approaches them and lies down. This lets the mongooses know the warthog wants to be cleaned.

The mongooses then run forward and crawl all over the warthog. Up to 20 mongooses may climb over the warthog. They happily eat the warthog's ticks and other parasites. The mongooses get a tasty meal, and the warthog gets cleaned. Everyone wins!

Mongooses like to eat ticks that are filled with blood. Yuck!

FACT FINDER!

A relationship like the one between warthogs and mongooses requires trust. That's because warthogs could kill mongooses with their tusks if they wanted to.

WHY WARTHOGS AND MONGOOSES ARE SPECIAL

Nature is full of mutualistic relationships. For example, bees get a meal of nectar from flowers. In return, bees carry pollen to other flowers. But the relationship between warthogs and banded mongooses is unusual.

It's uncommon to see a mutualistic relationship between two mammal species. Such relationships do occur sometimes. But in most cases, one of the mammals is usually a primate. The relationship between warthogs and mongooses is the first one known where neither animal is a primate. That makes it truly special!

Scientists don't know if warthogs and mongooses naturally know how to act this way together or if they must learn how.

FACT FINDER!

A primate is any animal from the group that includes humans, apes, and monkeys.

WARTHOGS AND BIRDS

Warthogs enjoy an unusual and special relationship with mongooses. But this isn't their only mutualistic relationship with another creature. They have also partnered up with some birds.

Small birds called oxpeckers ride along on warthogs and eat parasites off them. This relationship isn't surprising. Oxpeckers have the same relationship with many large African mammals. But scientists were surprised to discover warthogs also have a mutualistic relationship with birds called southern ground hornbills. These are turkey-sized birds with large, powerful bills!

warthog and oxpecker ▶

Oxpeckers may help warthogs, but they might also harm them. The birds sometimes eat blood from wounds on the warthogs, which causes the wounds to take longer to heal.

FACT FINDER!

Southern ground hornbills use their bill to gently remove parasites from warthogs. But that bill is strong enough to kill a tortoise!

WARTHOGS AND TURTLES

A few years ago, scientists happened upon a highly unusual scene. A warthog went into a **water hole**—and then two turtles swam over and started eating parasites off it!

This kind of event had never been photographed before. The warthog stayed still even while the turtles picked ticks off its face. The large mammal then dropped down lower into the water so one turtle could grab a big fly off its back. It seems the warthog and turtles all benefited from their water hole meeting!

South African helmeted turtle

From turtles to mongooses, warthogs have partnered up with many different animals to solve their parasite problems!

GLOSSARY

burrow: a hole made by an animal in which it lives or hides

flea: a very small bug that lives on animals and sucks blood from them

mammal: a warm-blooded animal that has a backbone and hair, breathes air, and feeds milk to its young

parasite: a living thing that lives in, on, or with another living thing and often harms it

protect: to keep safe

relationship: a connection between two living things

tusk: a large tooth that curves up and out of an animal's mouth

wart: a small, hard bump caused by a virus

water hole: a small body of water used by animals for drinking or cooling off

FOR MORE INFORMATION

Books

Furstinger, Nancy. *Warthogs*. Lake Elmo, MN: Focus Readers, 2018.

Hansen, Grace. *Warthog*. Minneapolis, MN: ABDO Kids, 2018.

Schuetz, Kari. *Warthogs and Banded Mongooses*. Minneapolis, MN: Bellwether Media, 2019.

Websites

Banded Mongoose
www.folly-farm.co.uk/zoo/meet-the-zoo-animals/banded-mongoose/
Discover more about banded mongooses here.

Do Warthogs Really Have Warts?
wonderopolis.org/wonder/do-warthogs-really-have-warts
Learn more about warthogs on this website.

Unlikely Partners: Warthog and Mongoose
ny.pbslearningmedia.org/resource/nat35-sci-unlikely-partners/unlikely-partners-warthog-and-mongoose/
Watch a great video of warthogs getting cleaned by banded mongooses here.

Publisher's note to educators and parents: Our editors have carefully reviewed these websites to ensure that they are suitable for students. Many websites change frequently, however, and we cannot guarantee that a site's future contents will continue to meet our high standards of quality and educational value. Be advised that students should be closely supervised whenever they access the internet.

INDEX

Africa, 5, 7, 8, 9

burrows, 6

carnivores, 4

fleas, 10, 12

herbivores, 4

humans, 10

mud, 12, 13

mutualistic relationship, 4, 16, 18

oxpeckers, 18, 19

parasites, 10, 12, 14, 18, 19, 20, 21

predators, 6

primate, 16, 17

Sahara Desert, 5

southern ground hornbills, 18, 19

ticks, 10, 12, 14, 15, 20

tsetse flies, 10

turtles, 20, 21

tusks, 6, 15

Uganda, 14